Vanesa Lucia Bazan Brizuela

Gold Pyrometallurgy

Vanesa Lucia Bazan Brizuela

Gold Pyrometallurgy

The merger process

ScienciaScripts

Cover image: www.ingimage.com

This book is a translation from the original published under ISBN 978-613-9-43709-2.

Publisher:
Sciencia Scripts
is a trademark of
Dodo Books Indian Ocean Ltd. and OmniScriptum S.R.L publishing group

120 High Road, East Finchley, London, N2 9ED, United Kingdom
Str. Armeneasca 28/1, office 1, Chisinau MD-2012, Republic of Moldova, Europe
Printed at: see last page
ISBN: 978-620-8-25189-5

INDEX

CHAPTER:1- GENERAL CONCEPTS 3

1.1 INTRODUCTION 3

1.1 FUNDAMENTAL PRINCIPLES OF EXTRACTIVE METALLURGY. 4

1.3Extractive metallurgy procedures. 5

CHAPTER:2- PYROMETALLURGICAL GOLD PROCESSING 8

2- PYROMETALLURGICAL PROCESS 8

2-1 Retorta Furnaces 8

2-1.1 Distillation of Mercury: 10

2.1.2 Distillation in Retorts: 10

2.1.3 Mercury Storage: 12

2.1.4 Mercury Toxicity: 12

Chapter 3- Gold Smelting 13

3-1GOLD fusion 13

3-1-1 Flame Color 15

3-2 OVENS: 16

3-2-1Induction Furnace 19

3-3-FURNACE OPERATION. 19

3-4-DORE FUSION 20

3-5FUNDING OF THE LOAD: 21

3-6-Extraction of noble metals from ores 23

Chapter 4- Fluxes. 25

4-1-1Limestone, iron oxides and silica. 25

4-2Oxidizing fluxes. 25

4-3Reducing fluxes. 26

4-4 Neutral fluxes. 26

4-5Effect of other constituents. 26

4-6-DOSING OF FLUXES 30

4-7-Fusion of the load 31

Chapter 5- PHASES 34

5-1Slag phase. 34

5-2 METAL PHASE. 34

5-3 Phase Mate. 34

5-4 Phase SPEISS. 35

5-5-Doré 36

5-6- Slags 37

5-6-1-Characteristics of the slags 38

5-6-2Physical Properties of Slag. 39

5-6-2Chemical Properties of Slag. 41

Chapter 6- Refractories.. 45

6-1Classification of Refractories... 45

6-2Refractory in gold production furnaces 46

6-2-1Crucible performance. 47

6-2-1Crucible condition repair and replacement 48

6-3 WEAR OF Refractories... 48

6-3-1Corrosion and wear mechanism of refractories.. 49

Bibliography... 50

CHAPTER:1- GENERAL CONCEPTS

1.1 INTRODUCTION

The high temperature process is widely used in the production of metals and materials. There are several reasons for using it:
The relative stability of metals and their components changes considerably with temperature, making it possible to achieve chemical and structural changes in the different phases present in the system. The rates of mass transport and chemical reactions increased with the increased thermal energy allowing changes to be achieved in less time.
The liquid and gas phase process which is possible at high temperatures not only allows reactions to take place at a faster rate but also allows phase separation to be achieved with relative ease.

There are no conventional limits for the temperatures used in the high temperature process.

In the case of metals and materials processing, however, for the reasons mentioned above, most processes are carried out between 300 and 2000°C temperature. The term pyrometallurgical process is often used to describe high temperature processes related to metal production. However, the basic principles underlying the use of high temperatures are common to the processing of all materials. The following examples illustrate the range of applications of high temperature processing in the metals and materials industries.

Perhaps the simplest example of heat treatment is drying. Physical processing techniques cannot remove the last traces of physically absorbed and incorporated water from finely divided materials and heat is used to remove the remaining water from the gas phase as vapor. However, rapid heating of unfinished ceramic components would cause severe deformation and fracture, therefore, drying must be carried out under controlled conditions to avoid product losses.

Thermal decomposition of inorganic compounds, e.g. calcination of limestone, $CaCO_3$, is a very common practice. Limestone is a feed storage in several important chemical processes in industry and during heating it separates into lime, CaO and CO_2 . Metal hydroxides, sulfates and carbonates, obtained by chemical precipitation from aqueous solutions, are used in the production of ceramic materials. At high temperatures, these materials separate into their respective oxides. The components formed and their physical properties can be altered by controlling the composition of the starting material, the treatment temperatures and the atmosphere in which the high temperature process takes place.

Metal oxides, sulfide and halides are the major sources of metals and a wide variety of options for the treatment of these materials are available in the high temperature pyrometallurgical process.

Another interesting point to note is that combinations of high temperature, organic/aqueous solution and electrochemical techniques are often used to obtain the expected products. Where new components are obtained then the reader should refer to the alternatives for processing the new component, in the event that further processing is considered, for example, if the original material is a metal sulfide and this is transformed into an oxide, then the processing options for metal oxides should be examined.

1.1 FUNDAMENTAL PRINCIPLES OF EXTRACTIVE METALLURGY.

The first step in obtaining a metal is to discover the place where one of its minerals exists in adequate quantity, i.e. a deposit. This is a prospecting work carried out by geologists using different methods: mechanical, chemical, gravimetric, magnetic, electrical, aerophotographic, among others.

Once the deposit where the metalliferous mineral or ore exists is located, the necessary considerations are made to determine its economic exploitation.

The exploitation of the rock containing the ore is carried out by the mining engineer who can do open pit mining or mining with tunnels using various techniques.

The exploitation of the deposit is a mine from which the material is extracted and delivered to the extractive metallurgical engineer, who must follow at least three steps to convert it into metal:

Separation of the ore and gangue (which is the non-useful part), in an operation known as mineral beneficiation or mineralurgy.

Preliminary chemical treatment that produces a compound suitable for metal reduction.

Reduction to metal, possibly with a subsequent refining treatment.

To separate the metal from the gangue, the first step is usually crushing and grinding. Sorting follows, which may be by dry screening or sieving or by other means, in which the large pieces are separated from the fines. Sometimes it is necessary to re-agglomerate the fines to a suitable size, one method for agglomeration is sintering.

Concentration of the mineral, to enrich it for transport or processing, can be done gravimetrically using water currents (calones, Wilfley table, elutriators, among others) or air (cyclones), by electrical or magnetic separation, by flotation - in which one type of particle is floated by means of reagents while the others are sedimented -, by chemical reactions or by other methods depending on the type of mineral and the needs of the process.

The ore concentrate can be treated by calcination and roasting - where the sulfide metal, for example, is converted to oxide - by leaching or by other chemical or electrochemical processes.
he reduction of the prepared ore is carried out in furnaces and converters of different types when pyrometallurgy is used or by chemical or electrochemical means when hydrometallurgy is used. At the end of the refining process a metal or alloy is obtained.

Metallurgy Concept

Extractive metallurgy can be defined as that part of metallurgy that studies the chemical methods necessary to treat a mineral ore or a material to be recycled in such a way as to obtain, from either of them, the metal, more or less pure, or some of its components.

1.3EXTRACTIVE METALLURGY PROCEDURES.

Extractive metallurgical operations fall into one of two groups: dry process operations and wet process operations. The former are generally referred to as pyrometallurgical operations and the latter as hydrometallurgical operations. Dry process operations are carried out at high temperatures between products in solid, liquid or gaseous state, while wet process operations are carried out through reactions in aqueous phase and at low temperatures.

Both pyrometallurgy and hydrometallurgy can be divided according to the possible operations that can be performed. The following table sets out this division.

Extractive metallurgy division

PYROMETALLURGY	HYDROMETALLURGY
-Calcination -Oxidizing Caking Oxidizing Sulfating Sulfating Chlorinating Binder Other -Ultrareductant Neutral Oxidizing Ultrareductant Neutral Reducing Fusion -Oxidant Reducing volatilization Of halides Of carbonyls -Igneous electrolysis -Metallothermy	

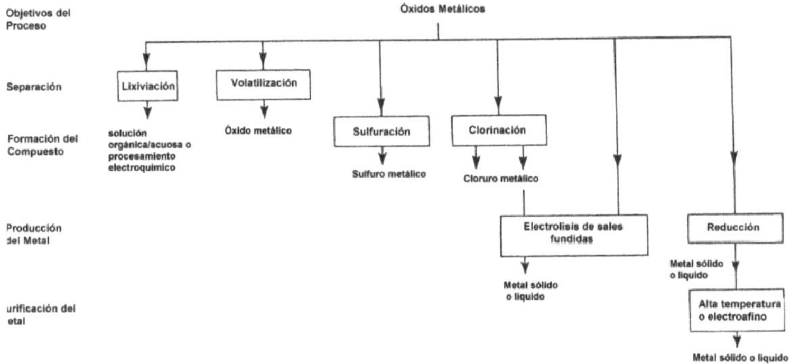

Figure 1. Alternative process routes for the treatment of metallic oxides

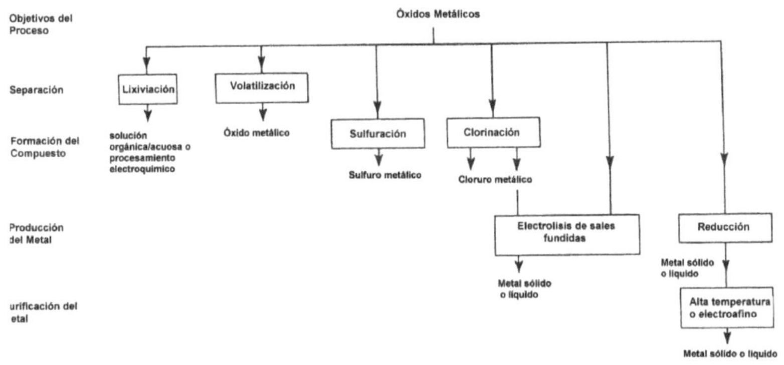

Figure 2. Alternative process routes for the treatment of metallic sulphides

2- PYROMETALLURGICAL PROCESS

The process begins with the collection of the precipitate from the Process plant, which is retained in three Filter Presses. The filtered solution, which is called Barren Solution and contains less than 0.02 ppm Au and Ag, is received in a tank and then pumped to the Leach Pad for irrigation of the heaps. The retained solid is collected every 6 to 7 days, depending on the amount precipitated, and is received in trays. This precipitate has a humidity of 35% and an average content of 25% Au, 57% Ag and 10% Hg.

Then, the precipitate is transferred to four Retort Furnaces. The purpose of these is to dry the collected precipitate and recover all the Mercury that is in it, for that reason it is worked with temperature ramps until reaching a maximum of 550 °C. The total cycle of the Retort is 24 hours and it works under a vacuum condition of 7" Hg. The removed Mercury is collected by a system of water cooled condensers and stored in a collector which is discharged at the end of the cycle to special Hg containers (flasks) for safe storage.

In order to remove any remaining mercury gas that may go into the environment, the vacuum stream passes through a water-cooled aftercooler located immediately downstream of the collector. This stream then passes through activated carbon columns and a water separator before going to the vacuum pump and is then discharged to the atmosphere. The saturation of the coals is controlled by constant monitoring. Mercury recovery is above 99%.

The dry and cold precipitate is mixed with the necessary fluxes and charged into two Induction Furnaces. It takes about 2 hours for the charge to melt completely and reach a temperature of 1300° C (approx.) in order to perform the slagging and the final casting to obtain the Doré bars.

2-1 RETORTA FURNACES

Electrical power is supplied to the heating elements (resistors) through the SCR units. The Retorta oven contains a total of 24 resistors. The sensor for the temperature controller is located at the rear of the Retorta.

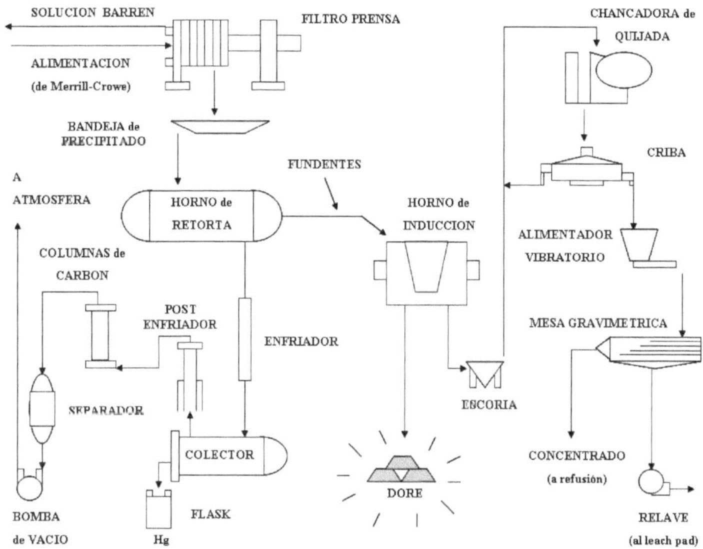

Figure 3: Process Flow Diagram.

The Retort Oven also has a high temperature controller. When a high temperature condition occurs, the high temperature limit sensor and lowers the SCR output to zero. A high temperature leakage sensor is located at the output and has an external alarm.

The Recorder has a swicth that allows switching from ON to OFF if desired.

The Honeywell remote bulb temperature controller controls the exhaust fan. This system controls the temperature limits of the air flowing through the extractor to atmosphere to prevent premature wear on the drive belts from the motor to the impeller. This value should be set to a maximum of 210°F (99°C).

A process timer causes the SCR output to turn off after the set time is met. In addition, a sensor is used to by-pass the temperature bulb of the controller. During the heating cycle, the exhaust fan is fully open and directed to the Retort door. After the operating cycle, the exhaust valve is reversed by the "TC1" controller and causes the airflow to be directed to pass through the heaters, facilitating their cooling.

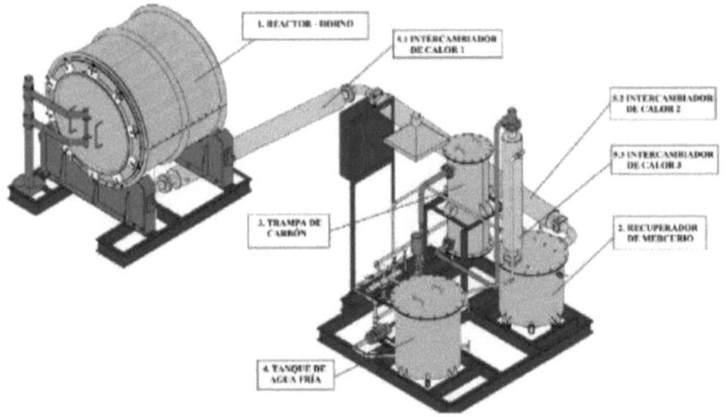

Figure 4: Retort kiln.

2-1.1 DISTILLATION OF MERCURY:

Depending on the type of ore to be processed (mineralogy) some contain significant concentrations of Mercury (>0.1 - 0.5%) and must be treated to remove them prior to smelting of the precipitate. This treatment must be done to minimize the release of toxic Mercury gases into the atmosphere during the next stages of the process.

Because of its high relative vapor pressure (1.3 x 10^{-3} mm Hg at 20°C) compared to other metals (Au = 10^{-10} , Ag = 10^{-22} and Pt < 10^{-22} mm Hg), mercury can be efficiently separated from other precious metals or bases by simple distillation. Mercury is removed by Retorts, furnaces specially designed for this purpose. Temperatures are similar to those applied for roasting or calcination. Other reactions occurring under these conditions are also applied during retorting.

The temperature of the Retort is slowly increased to dry it completely before vaporizing the Mercury and to allow time for the Mercury to migrate to the surface. The system is maintained at maximum temperature for 10 hours to ensure complete volatilization of the Mercury. Removals of 99% are easily obtained.

2.1.2 DISTILLATION IN RETORTS:

Retorts are operated under slight negative pressure and the Mercury vapor is usually recovered within a countercurrent water condensation system. The vapor is rapidly cooled to below boiling point (356°C) and the liquid Mercury is collected under water to avoid re-evaporation.

Mercury losses in the order of 0.2% or 0.4% are obtained per distillation cycle, these losses are generally the result of non-condensed mercury.

Pure Mercury normally boils at 356°C. However, the Mercury present in the precipitate is replacing atoms in the Gold structure, and this boiling point increases to 480°C as a result of a low Mercury concentration; fortunately, the boiling point of Mercury can be lowered by reducing the pressure in the system and can be improved by placing the precipitate in a vacuum system. Thus a vacuum pump is provided to reduce the pressure in the Retort to below atmospheric pressure.

The treatment that the precipitate (pptdo) receives in the Retorts is a roasting with temperature ramps that help us to extract as much Mercury as possible from the material. The ramps make up a 24-hour working cycle of the Retorts, which can be seen in Figure 5.

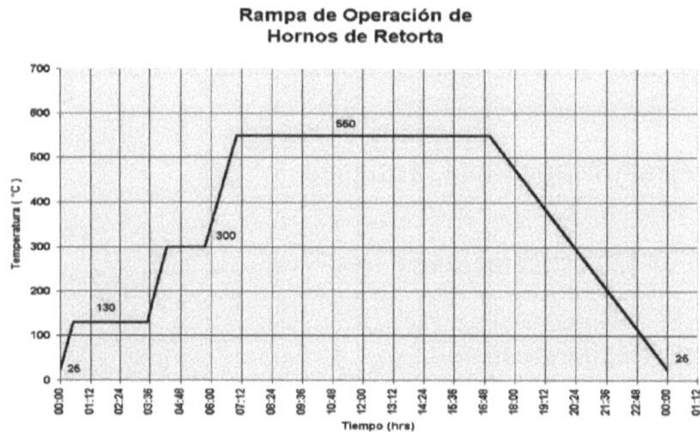

Figure 5: Retort Furnace Operation Ramp.

The Retort uses water-cooled condensers for Mercury condensation. The collector tank temporarily stores the Mercury. The aftercooler is water-cooled and condenses any remaining Mercury. Four columns of Retort activated carbon are located at the suction of the vacuum pump. A vacuum pump is installed to create the necessary vacuum in the Retort.

2.1.3 MERCURY STORAGE:

Once the Retort cycle has been completed, the recovered Mercury is drained from the collecting tanks into steel bottles that are manufactured from 3/8" thick steel plates. These bottles (known as flasks) are recyclable and reusable, the procedures and materials used for their manufacture meet American (EPA) and United Nations (UN) standards.

These Mercury flasks are temporarily stored within the Refinery area until they are shipped (exported).

2.1.4 MERCURY TOXICITY:

Mercury is highly toxic and has a cumulative physiological effect. If there is not good vapor control when treating precipitates with high Hg content, there can be serious problems. These effects can be counteracted by good operating efficiency as well as good hygienic and cleaning practices.

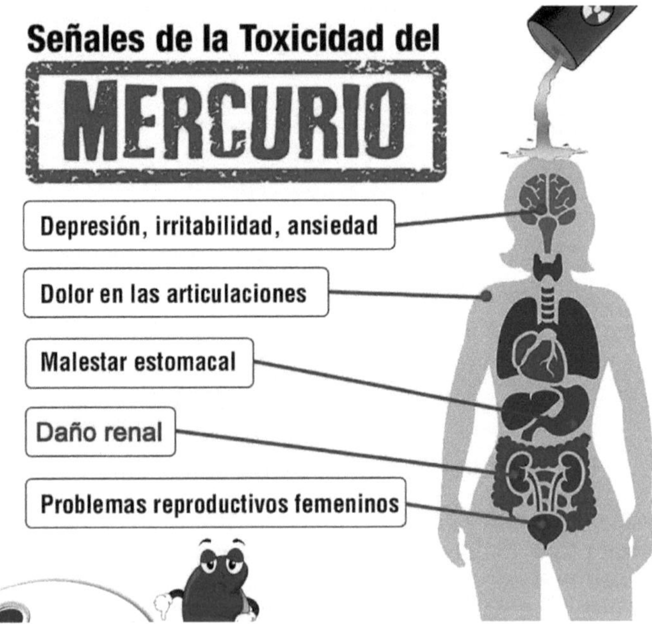

Figure 6: Effects of Mercury.

CHAPTER 3- GOLD SMELTING.

Smelting is a process that involves more than simply melting the metal to extract it from the ore. Most mineral ores are compounds in which the metal is combined with oxygen (in oxides), sulfur (in sulfides) or carbon and oxygen (in carbonates), among others. To obtain the metal in its elemental form, a chemical reduction reaction must take place to decompose these compounds. This is why smelting requires the use of reducing substances that, when reacting with the oxidized metallic elements, transform them into their metallic forms.

3-1GOLD FUSION

The cold and dry gold and silver product must be mixed with the necessary fluxes to charge the furnaces and proceed with the melting process.

It takes about 2 hours for the charge to melt completely and reach a temperature of 1100°C (approx.) in order to carry out the slagging and the final casting to obtain the Doré bars. The cascade casting system is used to obtain the bars.

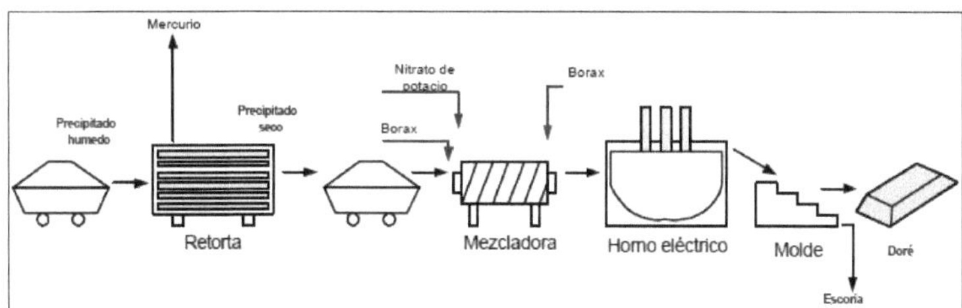

Figure 7: GOLD Smelting Scheme

- Gold is a precious metal that has a melting point of 1064°C. At atmospheric pressure,
- At temperatures above the melting point, gold volatilizes as red vapors. This volatilization increases with increasing temperature.

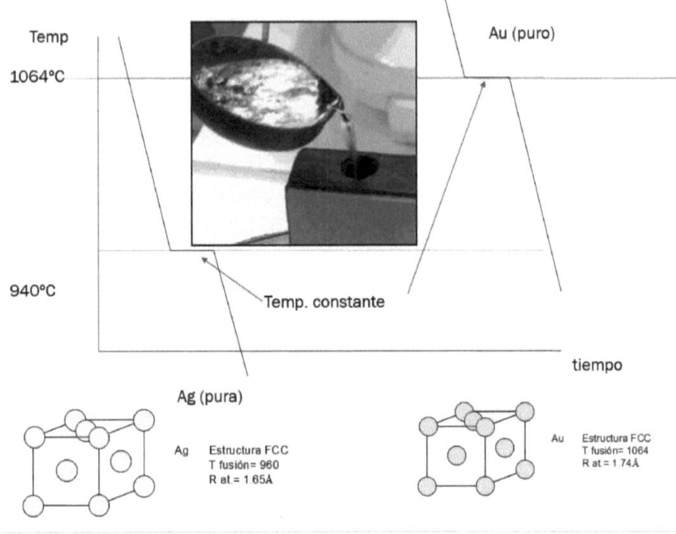

Figure 8- Au-Ag Binary Diagram

- At 1050°C volatilization is imperceptible and is still low below 1250°C. Volatilization increases with the presence of metallic impurities, particularly tellurium: for example, an alloy containing 5% Te loses between 2% and 4% of the Au content in one hour at 1245°C. Alloys containing 5% Hg or Sb lose approximately 0.2% Au under similar conditions.

Metal / Mineral	Punto de Fusión °C	Punto de Ebullición °C
Au	1064	2808
Ag	961	2210
Pt	1769	4530
Hg	-38.9	357
Zn	420	907
Pb	327	1744
Cu	1083	2595
SiO_2	1723	2230
$Na_2B_4O_7.10H_2O$	750	--
Na_2CO_3	851	--
$NaNO_3$	271	--
CaF_2	1403	2500
ZnO	1975	--
Ag_2O	230	--
PbO	886	--
Al_2O_3	2072	2980
Fe_2O_3	1565	--

3-1-1 FLAME COLOR

The flames present in the fire are not of the same color, it all depends on the material that is burning. Just as there are different colors, there are also different types of flames that indicate the special characteristics that make them up.

Throughout time, humans have used fire in a variety of ways and forms that have allowed them to develop increasingly sophisticated procedures to make life more comfortable. Visible light is generated by temperatures high enough and fast enough to be able to observe the molecules in movement, which produces the flame, and it can be determined that there are different colors associated with the temperature developed inside the ovens.

Temperature is one of the fundamental elements that will determine the color and intensity of the flame.

	Color	Temperature
1	Visible dark red	470 °C
2	Blood color	530 °C
3	Dark red	570 °C
4	Dark cherry red	650 °C
5	Light cherry red	750 °C
6	Light red	850 °C
7	Orange	900 °C
8	Light orange	950 °C
9	Yellow	1.000 °C
10	Light yellow	1.100 °C
11	White	1.200 °C

3-2 OVENS:

Furnaces used to melt metals and their alloys vary greatly in capacity and size, ranging from small crucible furnaces containing a few kilograms of metal to open hearth furnaces up to 200 tons capacity. The types of furnaces used in a smelting process are:

- Crucible furnace (mobile, stationary and tilting).
- Electric oven.
- Induction oven.
- Electric arc furnace.
- Tilting furnace.

- Cupola furnace

The type of furnace used for a smelting process is determined by the following factors:

- ✓ The need to melt the alloy as quickly as possible and bring it to the required temperature.
- ✓ The need to maintain both the purity of the cargo and the accuracy of its composition.
- ✓ The required furnace output.
- ✓ The operating cost of the furnace.

Crucible Furnaces: In these furnaces the metal is melted, without coming into direct contact with the combustion gases and for this reason they are sometimes called indirectly heated furnaces, the crucible is generally heated and then charged.

Moving crucible furnace: the crucible is placed in the furnace which uses oil gas or pulverized coal to melt the metal charge, when the metal is melted, the crucible is lifted out of the furnace and used as a casting ladle.

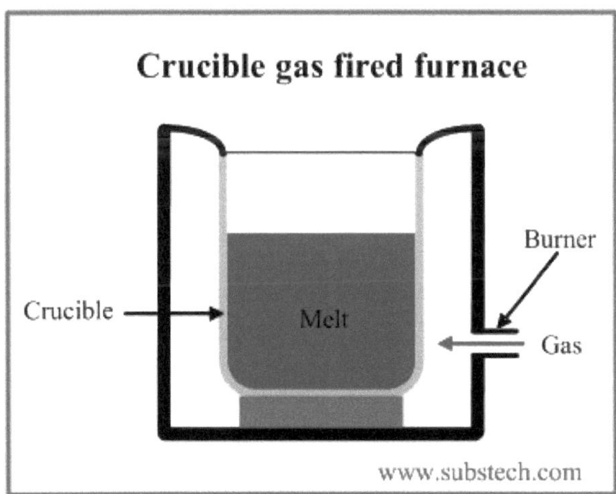

Figure 9: Mobile Crucible Furnace

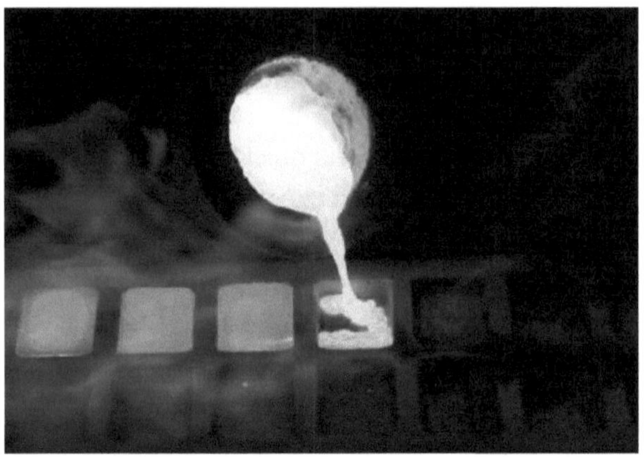

Figure 10: Crucible kiln in operation.

Stationary crucible furnace: in this case the crucible remains fixed and the molten metal is removed from the vessel by means of a ladle and then taken to the molds.

Tilting crucible furnace: the whole device can be tilted to empty the charge, they are used for non-ferrous metals such as bronze, brass, zinc and aluminum alloys, GOLD.

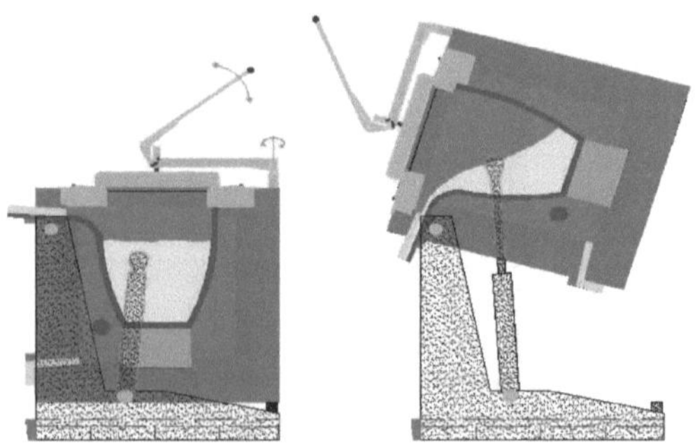

Figure 11: Induction Furnace.

3-2-1INDUCTION FURNACE

The induction melting furnace is a container formed by a helical circuit (coil) connected to an alternating current source and cooled with water. This coil is protected by refractory material, inside which is housed a removable Silicon Carbide crucible.

Kiln 12: Kiln Crucible.

3-3-FURNACE OPERATION.

The heat energy is achieved by the effect of the alternating current and the electromagnetic field that generate secondary currents in the load; the crucible is loaded with material, which can be scrap, ingots, returns, chips, concentrate or others.

When the metal is charged in the furnace, the electromagnetic field penetrates the charge and induces the current that melts it; once the charge is molten, the field and the induced current stir the metal, the stirring is a product of the frequency supplied by the power unit, the geometry of the coil, density, magnetic permeability and resistance of the molten metal.

Induction furnaces range in capacity from less than 1 kilogram to 320 tons and are used to melt all kinds of ferrous and non-ferrous metals, including precious metals.

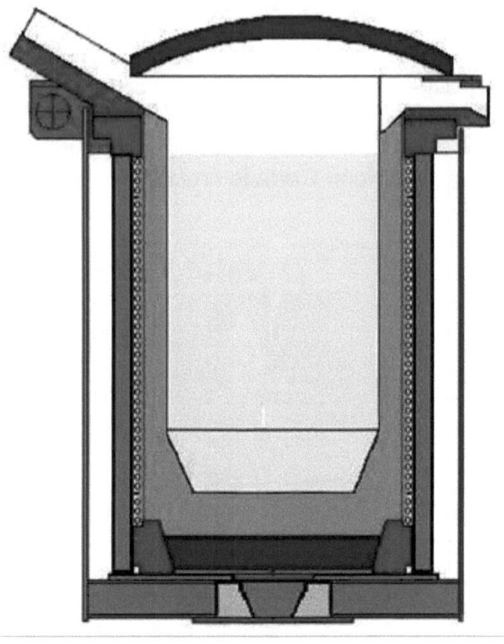

Figure 13: Schematic diagram of induction furnace operation

3-4-DORE FUSION

The final product obtained at the Refinery is Doré bars (also known as Bullion), which is an alloy of Gold and Silver. The melting point of the doré depends on the chemical composition of the alloy. Figure 2 shows the binary diagram for the Ag - Au alloy, which shows the different melting points for this alloy at different chemical compositions. Currently the Doré produced at the Refinery has a chemical composition of 19 - 21% Au and 78-80% Ag (on average). According to the aforementioned binary diagram, an Au - Ag alloy with that composition would have a melting point close to 980°C. This is more clearly seen in Figure 3. It is worth mentioning that this melting temperature is only for the Doré metal, not for the entire charge. This will be discussed later.

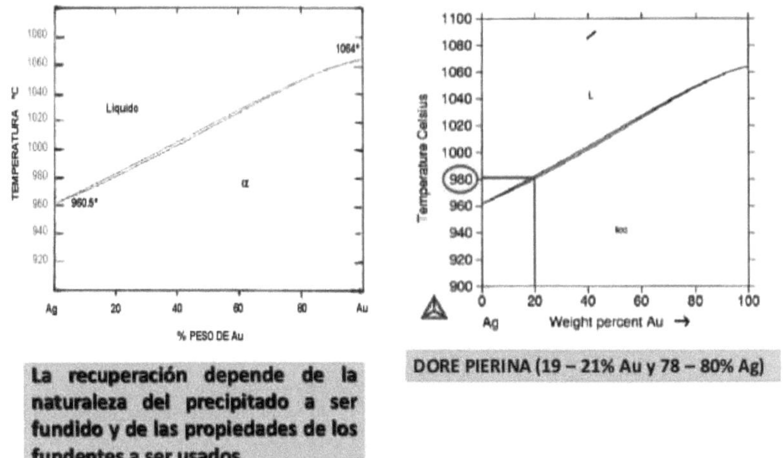

Figure 14: Binary diagram of Au-Ag

If the copper is not efficiently oxidized and removed in the slag, it remains in a metallic state and can form part of the doré, altering its melting point. A ternary alloy is then formed

3-5FUNDING OF THE LOAD:

Charge preparation is a critical task in the smelting operation. The precipitate and the material recovered from the slag are weighed and mixed with fluxes in appropriate proportions in order to obtain a slag with the following properties:

- Low melting point
- Low density
- Low viscosity
- High fluidity
- High solubility of the oxides of base metals
- Insolubility of precious metals
- Low refractory wear (corrosion / abrasion)
- Easy to break down for re-treatment

The efficiency of separation between the slag and the doré metal is measured in terms of Au and Ag grades in the slag or, in other words, the recovery of base metals (and other impurities) trapped in the slag. The performance depends on the nature of the precipitate to be smelted, based on its metal content and the properties of the fluxes to be used.

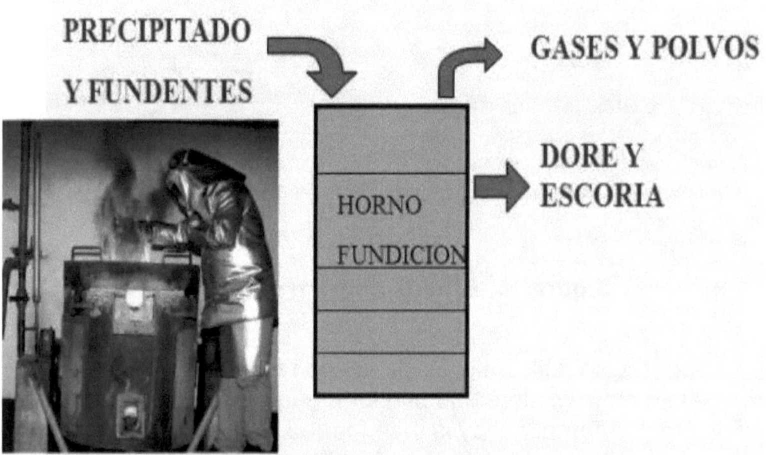

Figure 15: Mass balance

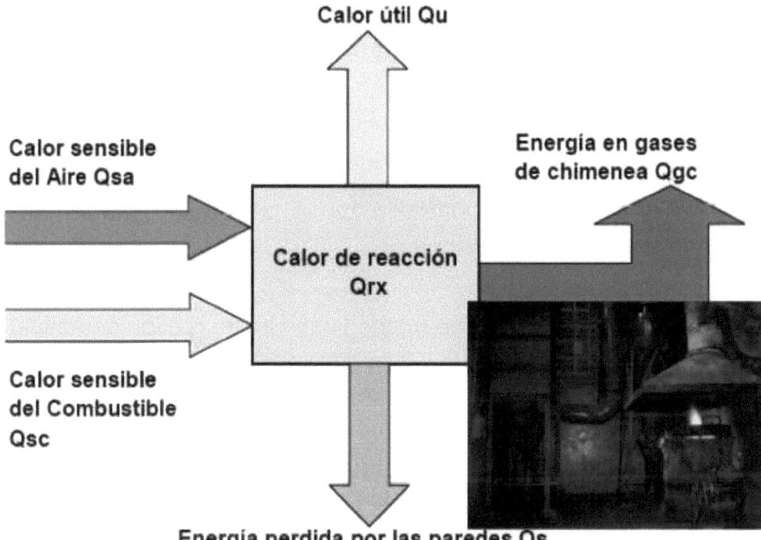

Figure 16: Energy balance

DORE PRODUCTION

3-6-EXTRACTION OF NOBLE METALS FROM ORES

The cyanidation process to extract gold (Au) and silver (Ag) from low grade ores, uses aqueous solutions of sodium cyanide (NaCN) with oxygen (O2), contained in the air, to convert the noble metal (M), from: .

M (solid with impurities) $M(CN)_2$ (soluble)

This requires an oxidizing agent, according to the following general equation known as the Elsner equation (Hedley and Tabachnick):

$$4Au + 8NaCN + O_2 + 2H_2O \Longrightarrow 4NaAu(CN)_2 + 4NaOH$$

Habashi reviewed the studies carried out on cyanidation mechanisms and proposed the equation for the dissolution reaction:

$$2Au + 4NaCN + O2 + 2H_2\,O \text{ ===> } 2NaAu(CN)_2 + 2NaOH + H\,O_{22}$$

Once brought into solution, the metals can be recovered from solution by:

- Precipitation, with Zinc (Zn) or Aluminum (Al) powder
- Adsorption on activated carbon
- Electrorecovery (Elecrowinning)

In the case of Petaquillas, the method used is Adsorption on Activated Carbon and subsequent Electrorecovery (EW).

The processing of the products from the electrowinning operation constitutes the final stage of production in the form of gold metal.

The objective of this stage is to purify the precipitates, removing the contaminants that accompany the precious metals and obtain doré metal in bars suitable for further refining. Common contaminants may include copper, lead, mercury, cadmium, and other metals.

The precipitates obtained by electrorecovery, apart from the contaminants coming from the solution, are accompanied by steel chips, which is the cathode material used in the electrolytic cells.

The cold and dry Gold and Silver product must be mixed with the necessary fluxes to charge the Furnaces and thus proceed to the melting process. It takes about 2 hours for the charge to melt completely and reach a temperature of 1200ºC (approx.) in order to carry out the slagging and the final casting to obtain the Doré bars. The cascade casting system is used to obtain the bars.

Gold is a precious metal that has a melting point of 1064ºC. At atmospheric pressure, Au bubbles at 2808°C and at 1800°C in a vacuum system. At temperatures above the melting point, gold volatilizes as red vapors. This volatilization increases with increasing temperature. At 1050ºC the volatilization is imperceptible and is still low below 1250ºC. Volatilization increases with the presence of metallic impurities, particularly tellurium: for example, an alloy containing 5% Te loses between 2% and 4% of the Au content in one hour at 1245ºC. Alloys containing 5% Hg or Sb lose approximately 0.2% Au under similar conditions.

CHAPTER 4- FLUXES.

The effect exerted on the melting point of one metal when alloyed with another is well known. An analogous effect exists among metal oxides. As a general rule, the melting points of these metal oxides are usually very high, much higher than that usually achieved in an industrial-type furnace. If one metal oxide is added to another, these small additions usually progressively lower the melting point of the resulting mixture, until the eutectic combination or lower melting point is reached.

It has already been pointed out that from an economic point of view only, it is necessary to use cheap materials (iron oxide, calcium carbonate and silica) as commercial fluxes.

The addition of sterile fluxes should be avoided if possible. If a flux is to be added, it is best that it be a basic or acidic material containing small amounts of beneficiable metals. Some ores or concentrates are called "self-fluxing" because they contain, in the right proportions, the elements necessary to produce a slag of the required silicate grade without resorting to the addition of a flux; however, these combinations are very rare.

4-1-1LIMESTONE, IRON OXIDES AND SILICA.

Silica is one of the most common and cheapest fluxes in the forms of sandstone, quartzite or quartz. It is often possible to use poor siliceous ores containing a small amount of gold and silver. However, many of these siliceous ores contain silicates of the feldspar, homablende and mica type, which do not contribute to the melting of the ore; on the contrary, they represent a burden for the homo which has to supply the heat necessary to melt them.

4-2OXIDIZING FLUXES.

This category includes:

Sodium and potassium nitrates,

Lead oxides and manganese,

Together with air as a source of oxygen.

4-3REDUCING FLUXES.

The only true reducing fluxes are cyanides. They are expensive and are only used with silver and gold ores in special processes.

4-4 NEUTRAL FLUXES.

Calcium fluoride is used because small amounts of this compound considerably lower the slag viscosity and melting point. This is also a very stable compound, which is the basic element of many molten baths used in electrolysis. (The double fluoride of sodium and aluminum, cryolite, possesses this same property, but is much more expensive).

Sodium sulfate is frequently used (as is fluoride) to dissolve adhering copper and lead slimes. It is also used (as a source of sodium sulfide) as a means of separating copper and nickel in the Oxford process.

Various chlorides are used as protective coatings for the molten ores, and not as solvents, since they are not very solvent-like.

Borax (Na B 0_{227} .10H_7 0) is also used for this purpose, although it is more expensive than ordinary chlorides.

4-5EFFECT OF OTHER CONSTITUENTS.

It has been pointed out that magnesium oxide is a common impurity in limestone and that when present in small quantities it tends to lower the melting temperature and specific gravity of the slag. Moreover, considering that its molecular weight is 40 and that of lime is 56, a lower weight of magnesium oxide guarantees the same degree of basicity in a slag, as well as being a cheaper flux and producing less slag. It is for this reason that some metallurgists prefer to use dolomitic limestone.

When the amount is large (above 5 percent) the magnesia tends to form a viscous slag with a higher melting point.

Fortunately, the presence of barium oxide is not common, as it is an undesirable impurity. It has a high molecular weight, gives a high density slag, and on melting of the matte forms barium sulfide which penetrates the matte and tends to make the slag lighter. As a result, the specific weights of the matte and the slag are very similar, which makes it difficult to separate them.

Zinc is also a very annoying impurity. In the melting of matte, it forms a sulfide that becomes part of the matte to dilute it and reduce its weight. In lead smelting, zinc tends to form complex silicates, which, due to their very high melting point, give rise to the formation of tufts in the crucible and on the walls of the high furnace, which causes many difficulties in the process.

The general effects of manganese oxide are similar to those due to iron oxide, although it produces a much less fusible slag than that formed with equivalent amounts of ferrous oxide and reduces the dissolving power of the slag with respect to zinc. In the lead furnace, the loss of silver in the slag increases.

The most commonly used fluxes are briefly described below:

- Borax: Sodium borate ($Na_2 B O_{47}$ .10 $H_2 O$), is an excellent base metal solvent. It has acidic characteristics and when molten dissolves practically all base metal oxides. Borax melts at 750°C, which lowers the melting point for all slags. It is very fluid when melted. Density: 2370 kg/m3.

Large amounts of borax can be detrimental by causing a hard and inhomogeneous slag. In addition, an excess of the reagent can hinder phase separation due to the reduction of the expansion coefficient of the slag and its action to prevent crystallization.

Borax dissolves most metal oxides so the purpose of borax in the flux mixture is to aid in slag formation. In small quantities, it lowers the temperature for slag formation and generates an orderly and quiet melt.

The dissolution of metal oxides by means of borax is carried out in two stages: first the borax is melted to a transparent glassy form, consisting of a mixture of sodium metaborate and boric anhydride, as seen in the reaction:

$$Na\ B\ O_{247} \rightarrow Na\ B\ O_{224} + B\ O_{23}$$

Boric anhydride reacts with the metal oxide to form a metal borate as shown in the reaction:

$$MO + B\ O_{23} \rightarrow MO.\ B\ O_{23}$$

There are five known classes of borates, which are classified similarly to silicates.

Nombre	Formula	Relación de Oxígeno Ácido : Base
Ortoborato	$3MO{\cdot}B_2O_3$	1:1
Piroborato	$2MO{\cdot}B_2O_3$	1.5:1
Sesquiborato	$3MO{\cdot}2B_2O_3$	2:1
Metaborato	$MO{\cdot}B_2O_3$	3:1
Tetraborato	$2MO{\cdot}2B_2O_3$	6:1

- Silica: Silicon dioxide (SiO_2) is added to the charge to balance the basic (caustic) content of the slag and produce a borosilicate slag. Pure silica melts at 1750°C and is the most available acidic reagent for melting. Silica-based slags are viscous and trap a lot of valuable metal in suspension. When silica is mixed with borax it forms a very fluid slag that can dissolve base metal oxides and combines with them in the form of stable silicates. Density: 2334 kg/m3.

- Sodium Nitrate: Density: 2260 kg.m-3. Sodium nitrate (NaNO3) is added to oxidize the base metals in the charge. This is a very powerful oxidizing agent whose melting point is 270ºC. At low temperatures nitro melts with little alteration; but at temperatures above 380ºC it decomposes producing Oxygen.

It oxidizes sulfides and some metals including lead, iron and copper.

A word of caution when using nitro is to control the amount required as the release of oxygen is a vigorous reaction and in addition to overflowing the crucible, will cause excessive erosion of the crucible. Nitro reacts with graphite in accordance with this reaction.

$$4\ NaNO_3 + 5C \rightarrow 2\ Na_2\ CO_3 + 3CO_2 + 2N_2$$

The addition of nitro is kept to a minimum because the release of oxygen causes a vigorous foaming reaction and can cause overflow in the crucible. It can also oxidize the crucible reducing its life.

Chemically, the following decomposition reactions are produced

$$2NaNO_3 + \text{Calor} \rightarrow 2NaNO_2 + O_2$$

$$2NaNO_2 + \text{Calor} \rightarrow Na_2O + N_2O + O_2$$

With further heating, potassium nitrite decomposes, giving potassium oxide, nitrous oxide and one mole of oxygen.

According to these reactions 1 mole of nitrate (85 g) produces 1 mole of oxygen (32.0 g).

• Potassium nitrate: has the same function as sodium nitrate. It reacts chemically as follows.

We must first determine the amount of O_2 generated by 1 mole of saltpeter when heated to over 400 °C.

$$2KNO_3 + \text{Calor} \rightarrow 2KNO_2 + O_2$$

$$2KNO_2 + \text{Calor} \rightarrow K_2O + N_2O + O_2$$

• Sodium Carbonate: Also known as Soda Solvay, $Na_2 CO_3$ (source of $Na_2 O$, to partly replace the silicate MO). It is a powerful basic flux and is by far one of the cheapest fluxes available.

It combines with the silica in the concentrate to form sodium silicate, releasing CO_2, according to the following equation:

$$Na_2 CO_3 + SiO_2 = Na_2 SiO_3 + CO_2$$

Due to the ease with which sulfides and alkaline sulfates are formed, it acts as a desulfurizing and oxidizing agent and has a very high fluidity. The use of soda provides transparency to the slag, excessive use causes sticky slag and hygroscopies.

In addition, a strong soda-based scum reacts exothermically with water and skin contact can cause severe irritation.

• Calcium Fluoride: It is also known as Fluorspar (CaF2). This additive reduces the viscosity of the slag by substituting the silicon ions for fluoride ions within the borosilicate slag structure, which causes a reduction in the viscosity of the system. Density: 3180 kg/m3.

• Manganese dioxide: Manganese dioxide is an effective oxidizing agent, but more importantly, it has a high affinity for sulfur and, in fact, is widely used in the desulfurization of steels.

Therefore, if a concentrate is known to have a high sulfide content, then manganese dioxide as a fluxing agent is more beneficial.

• Carbon: Used when a reducing atmosphere is required on a load.

As a general fluxing agent for the treatment of concentrates, it would find little use.

However, it is included in the list of fluxes because of its obvious reducing capabilities. Carbon does not become active until a temperature of 500-600°C, but it can be used effectively in a high copper content ingot to reduce intense oxidation.

It has been used for this function, as a heavy oxide layer can cause sampling and analysis problems.

4-6-DOSING OF FLUXES

The amount of flux needed depends on the quality of the precipitate based on its metallic content (gold and silver content).

The calculation necessary to determine the appropriate composition of the flux is based on the ternary diagram Na_2O - B O_{23} - SiO_2 , for the formation of slag, which is only referential to estimate the melting point of the slag. In the ternary diagram we are interested in knowing the temperature at which a solid mixture of a given composition (flux) will start to melt; or the amount of basic flux that needs to be added to an acidic precipitate or mineral gangue to achieve a slag that melts at the lowest possible melting temperature.

The working area for slag formation at the Refinery is shown in Figure 17.

The diagram in Figure 17 shows the melting points for various compositions of these three compounds (Na_2O, B_2 O3 and SiO_2) and which is formed by the added fluxes.

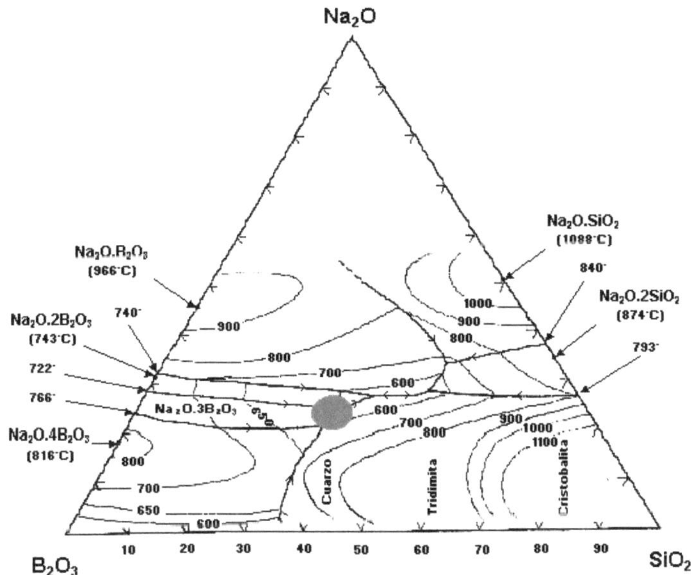

Figure 17. Ternary diagram of the Na system$_2$ O - B O_{23} - SiO_2

The theoretical dosage of fluxes to be added for both sulfide oxidation and oxide slagging is determined by considering the stoichiometry of the reactions that take place with each sulfide and metal oxide present in the ore or concentrate.

Once the sulfides have been oxidized, slags are formed, which consist of borates and silicates. The formation of metaborates and metasilicates is suggested because they are the most fluid and, being less viscous, allow the precious metals to pass more easily to the lower part of the crucible to form the metallic phase (Lenehan and Murray-Smith, 1986).

The oxidizing agents and fluxes for slag formation are mainly borax (Na B O_{247} .10H_2 O), and silica (SiO_2), which in suitable proportions react with the metal oxides to form metasilicates and metaborates. An example of this process is the slagification of iron oxide, which is illustrated in the reaction:

$$3Fe_2O_3 + 3Na_2B_4O_7.10H_2O + 6SiO_2 \rightarrow 3Na_2B_2O_4 + Fe_2(B_2O_4)_3 + 2Fe_2(SiO_3)_3$$

4-7-FUSION OF THE LOAD

The basic assumption for the retention of non-ferrous and ferrous metal impurities in the slag is their oxidation to form silicate or borate compounds. This assumption is thermodynamically feasible under the operating conditions found in the crucible.

The first chemical changes that occur during melting in the Induction Furnace are due to the effect of heat (Q), which causes the decomposition of the oxidizing fluxes, Carbonate and Sodium Nitrate:

$$Na_2\,CO_3 + Q = Na_2\,O + CO + \frac{1}{2}O_2 \;(851^{\circ}C)$$

$$NaNO_3 + Q = Na_2\,O + \frac{1}{2}N_2 + O_2 \;(308^{\circ}C)$$

In the presence of oxygen from the decomposition of the oxidizing fluxes, the oxidation of the base metals (which are in the charge as impurities) starts according to the reactions:

Determination of the amount of O2 generated by 1 mole of saltpeter when heated to above C400 °.

$$2\,KNO_3 + \varnothing \rightarrow 2\,KNO_2 + O_2 \quad O = 16{,}0$$

With further heating the potassium nitrite decomposes, giving potassium oxide, nitrous oxide and one mole of oxygen.

$$2\,KNO_2 + \varnothing \rightarrow K\,O_2 + N\,O_2 + O_2$$

According to these reactions 1 mole of nitrate (102.0 g) produces 1 mole of oxygen (32.0 g).

Then in the crucible, the nitrate releases oxygen to oxidize (or terminate the oxidation) of the sulfides and/or metals present.

$$Zn + \frac{1}{2}\,O_2 = ZnO$$

$$Pb + \frac{1}{2}\,O_2 = PbO$$

$$Pb + O_2 = PbO_2$$

$$Cu + \frac{1}{2}\,O_2 = CuO$$

$$2Cu + \frac{1}{2}\,O_2 = Cu\,O_2$$

$$Fe + \frac{1}{2}\,O_2 = FeO_2$$

$$4Fe + 3O_2 = 2Fe\,O_{23}$$

As a last stage, Borates and Silicates are formed by chemically reacting Borax and Silica with the oxides of the aforementioned base metals:

With Borax:

$$Na\ B\ O_{247}\ .10H_2\ O + Q = 2B\ O_{23} + Na_2\ O + 10\ H_2\ O\ (200^{\circ}C)$$

$$x\ Me_2\ O + y\ (B\ O_{23}\ ) = x\ Me_2\ O\ .\ y(B\ O\ )_{23}$$

With Silica:

$$x\ MeO + y\ SiO_2 = x\ MeO\ .\ y\ SiO_2$$

$$Me\ O_{23} + y\ SiO_2 = x\ Me\ O_{23}\ .\ y\ SiO_2$$

ME = METAL

CHAPTER 5- PHASES.

Having now discussed the main fluxing agents, the development of a flux recipe to suit a particular concentrate becomes essential. However, one must first be able to recognize the products resulting from the melting process. Basically, what results is a metal melt and a slag phase, but depending on the amount of contamination in the charge, the flux recipe or the ratio of flux to charge, there are two other possible phases: Mate and Spiess.

5-1SLAG PHASE.

They are solutions of oxides of different origins, as well as fluorides, chlorides, silicates, phosphates, borates, among others. The slag is the liquid with the lowest density and is therefore located in the upper part of the molten mixture.

The densities of some common slags are between 2.72 and 2.84 g/cm3 when at temperatures between 1823 and 1853 °K (Oliveira et al., 1999).

Slags are usually waste products and serve an important function, as they collect and remove most of the unwanted product from the valuable material. Another reason why slags are important is because they serve as thermal protection for the flux mixture, preventing heat loss from it.

5-2 METAL PHASE.

The metallic phase consists of pure metals, metal alloys, or solutions of non-metals with metals. In the liquid state, metals have low viscosities and high surface tensions, which results in a minimum contact angle between the liquid metal and the surfaces of the refractory materials that contain them, allowing them to flow more easily. The density of this phase is higher than that of the other phases, so it is found at the bottom of the molten mixture. For example, the density of gold is 19.3 g/cm3, of silver is 10.49 g/cm3, of copper is 8.96 g/cm3, of lead is 11.34 g/cm3, among others (. Generally, this is the valuable phase of the melting process and the one to be recovered.

5-3 PHASE MATE.

It is an artificial sulfide of one or more metals formed during the smelting process.

It is a very low melting point phase with a relatively high density and is therefore found between the metal and slag phases as a distinct and separate phase.

This layer is usually blue-gray in color, very brittle and may contain appreciable amounts of precious metals.

The mattes are usually iron or copper based and contain up to 10% gold in solution.

5-4 PHASE SPEISS.

It is an artificial metal antimonide or arsenide formed in smelting operations and is usually iron-based, however, cobalt and nickel can be substituted.

It is a hard, fairly tenacious, white tin substance that is found between the metal and slag layers, and is also capable of retaining an appreciable amount of precious metals in solution.

These two phases are mentioned because, as explained above, the formation of either phase is indicative of an incorrect flux formula or flux-to-charge ratio. Therefore, in most cases, a remelt with additional flux is sufficient to eliminate either phase.

However, if the flux recipes fail, then the matte can be broken down with nitro or manganese and can be broken down with caustic soda or sodium carbonate.

When treating material of this nature, it is usual to first melt the phase and then add the fluxing agent and stir into the molten charge. This increases the surface contact area and thus increases the reaction kinetics.

Again, some caution should be exercised when using nitro in this form, as the severe effervescence of the nitro as it decomposes may cause the material to overflow from the pot. Therefore, when using direct nitro, it is customary to use a large pot.

However, if the charge reacts turbulently, a small amount of dry salt will tend to calm the reaction. This is simply sprinkled on the surface of the slag.

Finally, when the reaction is complete, a quantity of borax equivalent to the flux used should be added. When fused, this is mixed with the charge before casting and will ensure that a workable slag is produced.

5-5-DORÉ

Doré is an alloy of Au and Ag. The objective of the smelting or melting process of Gold and Silver precipitates is to obtain Doré metal in the presence of slag-forming fluxes at temperatures exceeding the melting point of all the components of the charge typically between 1200 and 1300ºC. The time it takes to fully melt the charge depends not only on the quality of the slag formed but also on the chemical composition of the Gold-Silver alloy. The melting point of gold is 1064°C, while silver melts at 962°C. Figure 18 shows the Ag-Au binary diagram and it can be seen that the melting point of the alloy increases with increasing gold content.

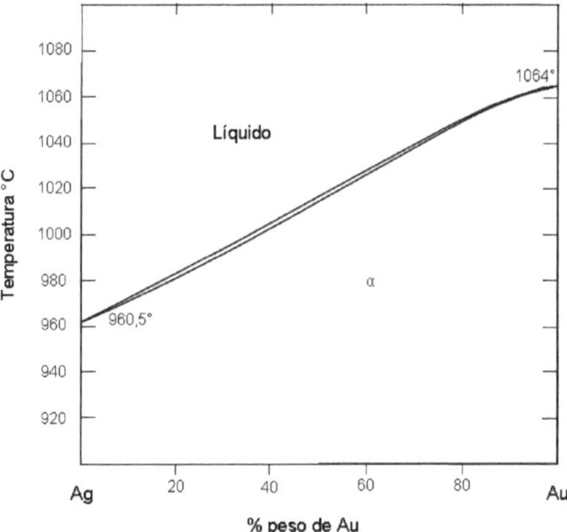

Figure 18. Ag-Au Binary Diagram

If the copper is not efficiently oxidized and removed in the slag, it remains in a metallic state and can form part of the doré, altering its melting point. A ternary alloy is then formed, as shown in Figure 19.

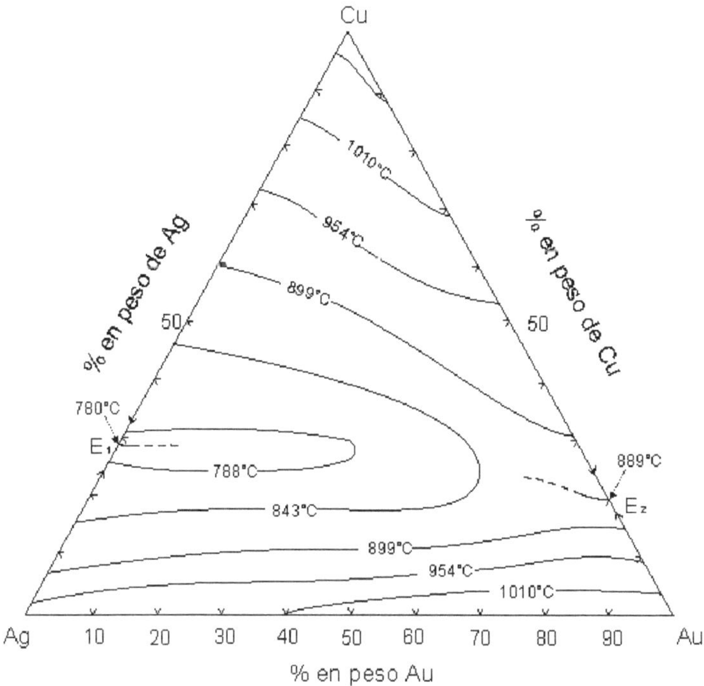

Figure 19. Ternary Diagram Ag-Au-Cu

Melting is a reduction or concentration process, in which most of the beneficiable constituents are collected in the form of metal, while the non-beneficial constituents form another product known as slag.

Roasting and calcination are usually preliminary processes carried out in order to obtain the metal in a more suitable form for melting or to remove some impurities that may unduly interfere with this process.

In a metallurgical melting process, the temperature produced and maintained is the result of the algebraic balance between the heat absorbed and generated and that transported to the products formed by the chemical reactions and physical changes that take place during the process.

The primary reaction may be followed or accompanied by several secondary reactions. The formation of compounds in the slag, metal or matte causes thermal changes.

5-6- SLAGS

Slag is the vitreous mass that remains as a residue when a precipitate is melted. During smelting, the slag forms a phase that separates from the doré and, due to its immiscibility and lower density, is placed on top of it, thus achieving the separation of both phases (Au density = 19.32 gr/cm3, slag density = 2.53 gr/cm3).

For the formation of slag it is necessary to use various fluxing reagents (known as flux). Fluxes are understood as any substance or compound that is purposely added to the charge in order to facilitate the melting of high melting point components such as those involved in gold smelting.

The addition of fluxes is done mainly for the following reasons:

- **Reduction of volatilization losses:** Fluxes reduce the melting point of the charge to a level below the temperature that could cause volatilization. The melt forms glassy slag layers that physically cover the metal during melting, reducing the potential for volatilizing elements in the metal layer.
- **Bath protection:** The formation of a slag layer isolates the molten metal bath from the atmosphere to avoid possible oxidation reactions with the atmosphere. Excessive heat losses are also avoided.
- **Collection of impurities:** The fluxes react chemically with the impurities contained in the precipitate. The impurities form chemical compounds with the fluxes that are soluble in the slag.

5-6-1-CHARACTERISTICS OF THE SLAGS

In general, a vitreous or glassy appearance is the trademark of an acceptable slag.

To ensure the fastest processing times, all flux ingredients should be finely divided and intimately mixed through the concentrate before charging into the furnace.

This increases the surface contact area which, in turn, greatly aids reaction kinetics by establishing many reaction zones in the melt.

The slag produced must meet the following general characteristics:

- Low melting point.
- Low viscosity.
- Low density.

- High fluidity.
- High solubility of base metal oxides.
- Non-solubility of Gold and Silver.
- Do not alter the metallic state of gold and silver.
- Good separation of Doré metal.
- Low refractory wear (due to corrosion and/or abrasion).
- Easy to break for re-treatment.

5-6-2PHYSICAL PROPERTIES OF SLAG.

✓ Energy or surface tension.

The surface tension of a slag is an important variable for the formation of the so-called foaming slags required in metal refining, which facilitate efficient transport of material by ensuring a large metal-slag contact surface.

✓ Density

The density of the slag is lower than the density of the metallic phase and this is of great importance as it supports phase separation by stratification in pyrometallurgical processes. In this way the oxidized phase or slag is superimposed on the rich metallic phase, thus allowing an adequate separation.

Experimental studies, developed by Mori and Suzuki, brought slag density estimates closer by relating them to temperature.

$$\rho(g/cm^3) = 5.00 - 0.0027*T - 0.50*CaO^{0.67} - 2.22*SiO_2^{0.4} + 0.70*FeO^{1.2} - 3.2*Fe_2O_3^{0.4}$$

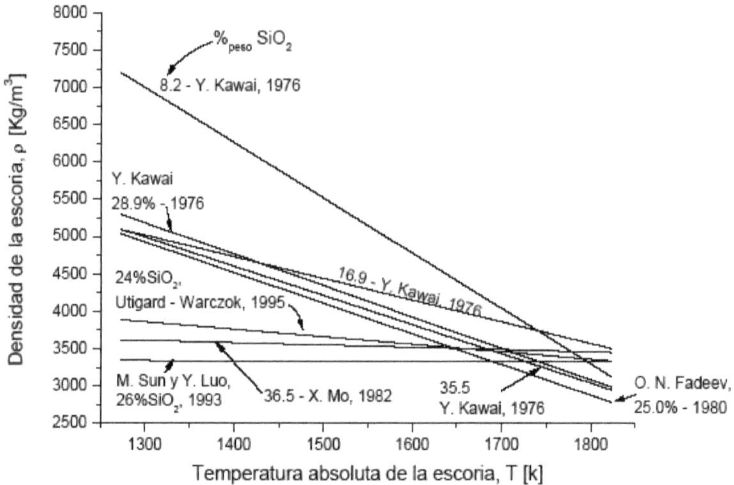

Figure 20- Density of slag

✓ Viscosity.

Slag viscosity represents one of the most relevant variables in most metallurgical processes and in the kinetics of conversion operations.

Several models have been published in order to estimate viscosity as a function of its chemical composition. Riboud's model and Urbain's model are some of the best known models for estimating viscosity. Riboud classifies the slag components into five categories, depending on their chemical nature and attributes values to the parameters of the Arrenihus-type equation according to the mole fraction of each category

$$\mu(T) = AT \exp\left(\frac{B}{T}\right)$$

The viscosity of the slag as a function of temperature is represented by the following mathematical equation:

$\mu = 1.336.10^{-4}$ e ^{14}

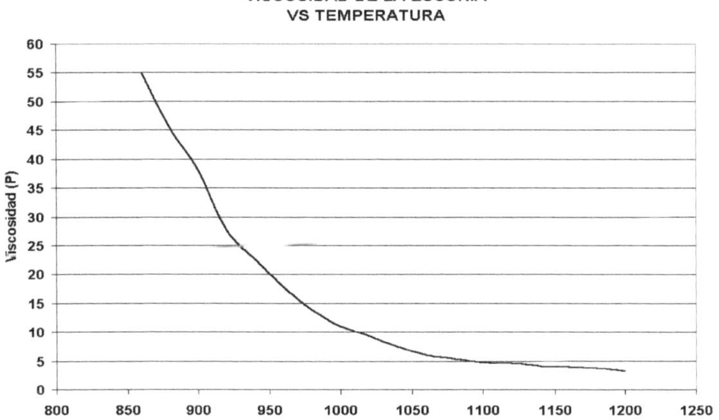

Figure 21. Relationship between Slag Viscosity and Temperature.

5-6-2CHEMICAL PROPERTIES OF SLAG.

Depending on the degree of acidity, the slag will be acidic or basic; depending on whether the slag components consume or release oxygen.

Oxygen consumers will be called acidic and those that give off oxygen in fusion will be called basic.

One of the ways to measure the level of acidity is given by the quotient given by **SiO_2 /(CaO+MgO+...).**

For an abundant acid gangue (SiO2, etc.), the slags must be basic.

However, if too much basicity is used to neutralize the acid gangue, slags that solidify at high temperatures may be obtained.

The molten slag should have a low melting temperature, otherwise, among the disadvantages derived from the lack of fluidity, there would be clogging during casting and mechanical entrapment of the alloy in the slag, which requires costly and cumbersome subsequent maneuvers.

It can be said that a material containing sufficient amount of silica will then be acidic.

$$SiO_2 + 2O^{-2} = SiO_4^{-4}$$

Classification.

As a general rule, the most common impurity to be separated is silica or some silicate; therefore, from a metallurgical point of view, silicate slags are the most important.

Many classifications of these slags have been suggested, but the one considered to be of importance is that which depends on the degree of silicate, that is, the ratio of oxygen in the acid to that in the base.

The slag formation and melting points must be low, otherwise energy consumption will be excessive and, sometimes, losses due to volatilization will also be high.

For the same reasons slags of high specific gravity are not desirable, since the matte or metal is not easily separated. Therefore, in industrial practice there should generally be a specific weight difference of at least one unit between the slag and the matte or metal.

The following table shows the different classes:

SLAG CLASSIFICATION

Name	Oxygen ratio from acid to oxygen of the base	Formula
Subsilicate	0.5 a1 ,0	**$4RO.SiO_2$**
Monosilicate	1,0 a1 ,0	**$2RO. SiO_2$**
Sexquisilicate	1,5 a1 .0	**$4R0.3SiO_2$**
Bisilicate	2,0 a1 ,0	**$RO. SiO_2$**
Trisilicate	3,0 a1 ,0	**2R0.3SiO2**

The following table can serve as a rough guide for the most common slags:

Specific weight

Monosilicates (Fe, Mn, Zn) 4,0
Bisilicates 3,5

Basic Silicates with Al 0_{23} .... 3,2 a 3,4 .

Acid silicates with Al 0_{23} ... ... 3,0 a 3,2

Mg Silicates ... ... 3,0 a 3,3

Ca Silicates ... ... .2,6 a 3,0

Silicates of Na, K 2,5

Barium Silicates 4,4

Lead Silicates 7,0

FeS 4,8

Cu_2 S 5,8

SiO_2 2,6

✓ Acid-base character

Slag basicity.

The basicity of slags has been expressed in various forms, but generally the essential expression of the basicity index is:

The binary (IB2), ternary (IB3) and quaternary (IB4) basicity index on the slag samples according to expressions (1), (2) and (3).

- IB2= CaO/SiO $_2$
- IB3=CaO/(SiO_2 + Al O $)_{23}$
- IB4=CaO + MgO/(SiO_2 + Al O $)_{23}$

The efficiency in the separation between the slag and the doré metal is measured in terms of Au and Ag grades in the slag or,

in other words, the recovery of base metals (and other impurities) trapped in the slag. The performance depends on the nature of the precipitate to be smelted, based on its metal content and the properties of the fluxes to be used.

CHAPTER 6- REFRACTORIES.

A refractory product is considered to be one that resists without melting or softening at temperatures equal to or higher than 1500 °C. for an economically profitable period of time, without excessive deterioration of its physical-chemical properties.

6-1CLASSIFICATION OF REFRACTORIES.

The classification can be made according to their chemical character, since this is what facilitates their choice according to their intended use. Based on this, they are classified as:

Acid refractories

- ✓ Clay products, such as refractory clay. The resulting refractory product after firing is composed of 54-73 % silica (SiO_2) and 20-45 % alumina (Al O_{23}). These products are quite sensitive to sudden changes in temperature.
- ✓ Silica products: quartz sand, quartzite. These products range from 100% silica (SiO_2) to products with no less than 85% silica and the rest alumina. They have a softening point of about 1400 °C, high resistance to alkali chloride vapors and high mechanical strength when hot. They are often used in the construction of coke ovens.

Basic refractories

- ✓ Aluminum oxides: They are mainly formed by silica and alumina, but unlike the acids, alumina predominates in these in proportions of 80-60 %. They have a softening point of around 1350 °C and generally have good properties to resist sudden temperature changes and abrasions.
- ✓ Calcium and magnesium oxides, such as dolomite [CaMg(CO $)_{32}$] and magnesia (MgO). The properties of both are very similar and dolomite can be mixed with magnesia in the form of blocks or paste, agglomerated with tar. The main element of magnesia refractories is periclase (MgO), characterized by its high melting point; 2800 °C, high thermal expansion, high refractoriness and high resistance to basic slags.
- ✓ To avoid volume variations at elevated temperatures, the raw material is sintered. In addition to MgO, sintered magnesia also contains varying amounts of Fe O_{23} , Al O_{23} , CaO and SiO_2 , which are important for the sintering process but lower the

softening point of the material and thus its usability. With small amounts of chromium, the resistance to sudden temperature changes is improved.

Neutral refractories

- ✓ Carbon products. These are prepared from graphite and coke. Graphite, agglomerated with clay, is used for the manufacture of crucibles with good heat conduction properties. Coke, agglomerated with tar and fired, is used for the construction of blast furnace hearths that resist the attack of melt and slag. However, they are sensitive to air oxidation and water vapor.
- ✓ Products based on mineral or artificial chromite, magnesia and lime. These materials are mainly composed of magnesia (MgO), lime (CaO) or both together with the chromium mineral; chromite ($Cr\ O_{23}$ -FeO), which also contains in its natural state an important portion of magnesium oxide (MgO) and alumina ($Al\ O_{23}$). The latter is quite chemically inert and highly resistant to both acidic and basic slags.

  Another material in this group is the chrome-magnesia refractory, consisting of a mixture of chromite and magnesia in a 1:2 ratio. The pieces are agglomerated by firing or by chemical bonding. The addition of chromium ($Cr\ O_{23}$) to the magnesia materials produces a high quality product with good resistance to sudden changes in temperature, fire and slag.
- ✓ Silicon carbides, zirconium, niobium, tantalum. Silicon carbide (SiC) melts at 2700°C in an oxygen-free atmosphere; in air, it oxidizes rapidly between 900-1300°C. It is used mixed with refractory clay to protect the silicon grains from oxidation. The resulting refractory product can have high coefficients of electrical and thermal conductivity, high mechanical strength at elevated temperatures and high abrasion resistance.

6-2REFRACTORY IN GOLD PRODUCTION FURNACES

The following refractories are used in the Induction Furnace:

Silicon Carbide Crucible: The main characteristic of this crucible is its chemical composition: 60% SiC and 30% C, which gives it acidic chemical characteristics. It is important to take this into account for a correct dosage of the fluxing reagents. This crucible also has excellent resistance to thermal shocks. The maximum working temperature is 1430°C.

The following refractories are used in the Induction Furnace:

Dri-Vibe Refractory: This refractory is used as a lining area of the top lid of inner lining of crucible furnaces . It is a fused alumina based material (86%) that has good resistance to chemical attack. The second main component is MgO with 8%.

The following refractories are used in the Induction Furnace:

Steel-Pak 86CR Refractory: This material was specifically designed for coreless induction furnaces. It is a high alumina material (76%) with phosphoric acid binder (4%) which gives it its plastic characteristics. The secondary components are SiO_2 with 13% and Cr O_{23} with 4%. The addition of chromite gives high corrosion and abrasion resistance to both the metal and the slag.

Minro-AL Plastic A-91: This material is used to manufacture the crown in the furnace. It is a high alumina material (87.4%) with phosphoric acid binder (3%) which gives it its plastic characteristics. The second main component is SiO_2 with 7%. It has good resistance to abrasion and corrosion, good workability, good adhesion and has good thermal resistance.

Refractory Minro Weave Cloth: This refractory cloth is used around the crucible lining to provide "relief" during the expansion of the crucible when in operation (melting). It is a textured fiberglass-based material capable of withstanding high temperatures in non-continuous operations. It is resistant to solvents and most acids and alkalis. It has a high dielectric strength and a low dielectric constant. It does not contain asbestos. It is presented in rolls.

6-2-1CRUCIBLE PERFORMANCE.

The performance of the crucibles has been evaluated according to the amount of precipitate and fluxes that can be processed per unit.

The amount of precipitate processed per crucible is analyzed, mainly due to the adequate addition of fluxes.

It has been seen that sodium nitrate is a strong oxidizing agent. If there is an excess of this component, a strongly oxidizing atmosphere is created and an accelerated "decarburization" of the crucible begins to occur, since the carbon contained in it begins to react directly with the sodium nitrate producing CO_2 and N_2 according to:

$$4NaNO_3 + 5C = 2Na_2CO_3 + 3CO_2 + 2N_2$$

This accelerates crucible wear and greatly affects its performance.

6-2-1CRUCIBLE CONDITION REPAIR AND REPLACEMENT

During melting, measure the temperature with the infrared thermometer in the furnace casing around the waist. If temperatures exceed 150°C, the crucible should be patched or replaced depending on wear and damaged areas.

1- Inject pressurized air (if the crucible is cold) to clean the residues that may remain after the casting and to have a better view of the state of the crucible.
2- Check the condition of the crucible as follows:
    a. Settling in for the best visualization of the crucible
    b. Observe in detail the depth of the scratches on the crucible wall.
    c. Observe the wear on the crucible wall (mostly at the waist) and on the floor caused by use.
    d. Determine the wear if it is greater than 60% of the thickness according to the design, proceed to patching.
3- If it is determined that the deterioration of the crucible due to wear, cracks, scour, etc. is greater than 60% of the thickness according to the design, the crucible is replaced or patched depending on the deterioration.

6-3 WEAR OF REFRACTORIES.

The corrosion of refractory materials by slag is a very complex issue due to the diversity of parallel mechanisms of chemical attack, thermomechanical and tribological wear processes that can influence during service, in addition to the fact that the physical and chemical characteristics during attack are not uniform due to the heterogeneous materials and corrosive media. As a consequence, both manufacturers and consumers of refractories must maintain continuous maintenance and research programs.

6-3-1CORROSION AND WEAR MECHANISM OF REFRACTORIES

✓ Chemical Factors

Chemical reactions between the brick components and the molten or gaseous metals penetrating through the porosity.

✓ Capillary Factors

Due to the open porosity of the refractory.

✓ Mechanical Factors

Erosion caused by kiln movement, agitation, impact of solid or liquid load and stresses resulting from expansion/contraction in masonries

Factor de Desgaste	Efecto	Propiedad Requerida
Químicos Escorias	Infiltración y corrosión de la micro-estructura Debilitamiento de la liga	Alta resistencia a corrosión Alta resistencia a infiltración
Capilares Porosidad abierta Características físico químicas del fundido	Infiltración en la micro-estructura Cambio de composición del refractario	Alta resistencia a la infiltración
Termo- mecánicos Infiltraciones metálicas Cambio de temperaturas Movimiento del baño/horno Tensión en la Mampostería Refractaria	Desgaste en el uso Aumento de la cond. térmica Desconchamiento (Spalling) Erosión Formación de micro-fisuras	Elevada refractariedad Alta resistencia a la inflitracion Flexibilidad estructural

BIBLIOGRAPHY.

1. Adamson R.J. 1972. Gold Metallurgy in South Africa. Chamber of Mines of South Africa, Johannesburg.
2. Coudurier, L., Hopkins, D. W., & Wilkomirsky, I. 2013. Fundamentals of Metallurgical Processes: International Series on Materials Science and Technology (Vol. 27). Elsevier.
3. Factory of ideas. 2014. Factory of ideas. Retrieved from The gold process from start to finish: https://fabricadeideas.pe/wp-content/uploads/2014/04/ProcesosYanacocha.pdf
4. García, e. Y. 2014. Design and construction of a crucible furnace for non-ferrous alloys. San Salvador: Universidad del Salvador.
5. Grimwade, Mark. 2000. "A plain's man guide to Alloy Phase Diagrams: Their use in Jewellery Manufacture".
6. Imris, Ivan. 2000. "Gold and Silver Smelting and Refining Processes."
7. Imris, Ivan. 2000. "Slag from Doré Smelting Process."
8. Integrated Global (n.d.). High Emissivity Coatings for Refractory Heater Surfaces. Retrieved from IGS: ttps://integratedglobal.com/en/services/high-emissivity-coatings-for-refractory-surfaces-heaters/#:~:text=The%20bricks%20refractory%20insulating%20(IFB,in%20the%20most%20efficient%20way%20).
9. John Marsden, Iain House.2006; The Chemistry of Gold Extraction; SME,;ISBN: 0873352408, 9780873352406
10. Kaspin, S., and N. Mohamad. 2015 "Gold refining process and its impact on the environment." In Environmental Engineering and Computer Application: Proceedings of the 2014 International Conference on Environmental Engineering and Computer Application , Hong Kong, 25-26 December 2014, p. 19. CRC .
11. Kaspin, Saadiah. 2013 "Small scale gold refining: Strengths and weaknesses." In Technology, Informatics, Management, Engineering, and Environment (TIME-E), International Conference on, pp. 32-36. IEEE, 2013.

12. Lenahan, W.e. and Murray-Smith R. 1986. Assay and Analytical Practice in the South African Mining Industry. Chamber of Mines of South Africa, Johannesburg.
13. Levin, Robbins and McMurdie .1969. Phase Diagrams for Ceramicists, 2nd edn. p.204. The American Ceramic Society.
14. McGuire, M.A. 1995. "Recovery of Precious Metals from Merrill-Crowe Precipitates by Smelting."
15. Palacios G, José. 1999. "Fundamentals of Fusion to Doré Metals".
16. Paredes, E. C. 2015. Optimization in the smelting of gold and silver precipitates. Perú.
17. Pillaca Hinostroza, I. 2017. Improvement of the fire assay process to determine Au and Ag in Pb-Zn and Cu concentrates in the company "MINLAB SRL"-Canaria.
18. Santander, Nelson H. 1997. "Refractory Materials".
19. Suzuki, M., Jak, E. Quasi-Chemical Viscosity Model for Fully Liquid Slag in the Al2O3-CaOO-MgO-SiO2 System-Part I: Revision of the Model. Metall Mater Trans B 44, 1435-1450 (2013). https://doi.org/10.1007/s11663-013-9928-3. https://doi.org/10.1007/s11663-013-9928-3.
20. Yáñez Concha, T. F. 2024. Optimization of the melting of gold gravimetric concentrates for Mina El Bronce.

Printed by Books on Demand GmbH, Norderstedt / Germany